Real-Life Math
DECIMALS AND PERCENTS

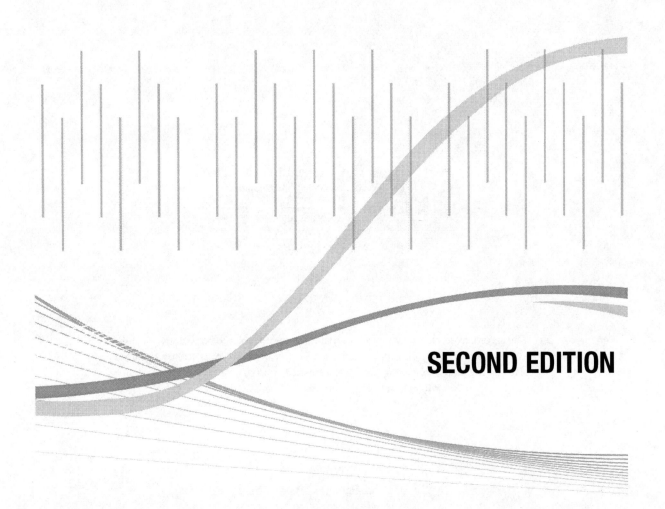

SECOND EDITION

WALCH PUBLISHING

SGS-SFI/COC-US09/5501

The classroom teacher may reproduce materials in this book for classroom use only.
The reproduction of any part for an entire school or school system is strictly prohibited.
No part of this publication may be transmitted, stored, or recorded in any form
without written permission from the publisher.

1 2 3 4 5 6 7 8 9 10

ISBN 978-0-8251-6325-8

Copyright © 1998, 2007

J. Weston Walch, Publisher

P. O. Box 658 • Portland, Maine 04104-0658

www.walch.com

Printed in the United States of America

Table of Contents

How to Use This Series .... *v*

Introduction .... *vi*

Reading and Writing Decimals and Percents

1. The Track Banquet 1
2. Cover Design 3
3. Tax Time 5
4. Finding Your Way 7

Decimals and Percent Equivalents

5. Swish 9
6. Doing Your Part 13
7. Who's Watching What? 15
8. Metric Measures 17

Comparing Decimal Numbers

9. The Batting Order 19
10. Fast-Food Wars 21
11. Which Costs Less: UPS or the U.S. Postal Service? 24
12. Dividing Up Lunch 27

Basic Operations with Decimals

13. The Building Challenge 29
14. Money Exchange 32
15. Scorekeeper 35
16. Movie Night 37

Percent of a Number

17. TV Time 40
18. Winning Percentage 43

Table of Contents

19. Planning a Trip 45
20. Deductions 47

Percent of Increase/Decrease

21. How Much Is School Worth? 50
22. Bigger, Stronger, and Faster 53
23. The Ad Campaign 56
24. CDs 58

Problem Solving with Decimals and Percents

25. GPA 60
26. Decisions, Decisions 63
27. Owner Satisfaction 65
28. On-Time Arrivals 67

More Problem Solving with Decimals and Percents

29. The Big Sale 70
30. Will It Ever Fit? 72
31. How Many Calories Can I Eat? 75
32. Bungee! 78

How to Use This Series

The *Real-Life Math* series is a collection of activities designed to put math into the context of real-world settings. This series contains math appropriate for pre-algebra students all the way up to pre-calculus students. Problems can be used as reminders of old skills in new contexts, as an opportunity to show how a particular skill is used, or as an enrichment activity for stronger students. Because this is a collection of reproducible activities, you may make as many copies of each activity as you wish.

Please be aware that this collection does not and cannot replace teacher supervision. Although formulas are often given on the student page, this does not replace teacher instruction on the subjects to be covered. Teaching notes include extension suggestions, some of which may involve the use of outside experts. If it is not possible to get these presenters to come to your classroom, it may be desirable to have individual students contact them.

We have found a significant number of real-world settings for this collection, but it is not a complete list. Let your imagination go, and use your own experience or the experience of your students to create similar opportunities for contextual study.

Introduction

Organization

The book is organized around four themes or contexts that are of high interest to students: Sports, Money, Entertainment, and Travel/Transportation. Within each context, there are eight different concepts or topics addressing decimals and percents. The concepts are Reading and Writing Decimals and Percents; Decimals and Percent Equivalents; Comparing Decimal Numbers; Basic Operations with Decimals; Percent of a Number; Percent of Increase/Decrease; Problem Solving with Decimals and Percents; and More Problem Solving with Decimals and Percents. The activities are grouped by concept, with four different contexts for teaching each concept. Choose the context—or contexts—that you find most appropriate for your students.

Order of Activities

The activities in the book are arranged to reflect the order in which decimal and percent concepts are presented in many textbooks. As such, you can supplement or enrich a concept presented in the textbook with this resource, or use the activities as an introduction to a new concept. The activities can also be done in any order; however, before students do the problem-solving and percent activities, they should have some facility with the concepts presented in the first part of the book.

Level of Difficulty

Some activities use more difficult mathematical concepts than others. As a general rule, the activities in the second half of the book are more difficult than those in the first half. It should be noted that the lessons that are less difficult mathematically still involve using higher-order thinking skills.

Time Considerations

Since student ability levels and school schedules vary greatly, time suggestions for the activities are not given. Before using an activity, review it and decide how much time would be appropriate for your students.

Calculators and Other Technology

A practical way of using calculators with the activities is to allow them if the situation described in the activity would warrant the use of a calculator in real life. In some of the activities, students can use spreadsheets, word processing, and desktop publishing software.

(continued)

Introduction

Organizing the Classroom

The Teacher Guide pages list suggestions on how to arrange students for the activities. Some of the lessons work best for individual student work, other lessons are more appropriate for students working in pairs, and some lessons work best for groups of students. The final decision on how to organize your students is left up to you.

Evaluation and Assessment

In cases where appropriate, selected answers are given. However, since the lessons model real-life situations, exact answers cannot always be provided.

Reading and Writing Decimals and Percents

teacher's page

1. The Track Banquet

Context

sports

Topic

reading and writing decimals and percents

Overview

In this activity, students assume the role of a track coach who is preparing a speech for an upcoming sports banquet.

Objectives

Students will be able to:

- write decimal numbers using words
- read decimal numbers out loud

Materials

- one copy of the Activity 1 handout for each student

Teaching Notes

- Students should complete questions 1 through 5 on their own. Once they are finished answering those questions, they can partner with another student and read their answers to each other out loud.

- Introduce the activity by eliciting from students when they think a coach might have to read or write decimal numbers.

- Before the activity, set the stage by explaining to students that they will be assuming the role of a track coach who is preparing to give a speech at a sports banquet.

Answers

1. ten and ninety-one hundredths; ten point nine one; twelve and one hundredths; twelve point zero one

2. forty-four and three hundred twelve thousandths; forty-four point three one two; fifty-one and eighty-three hundredths; fifty-one point eight three

3. fourteen and five hundredths; fourteen point zero five

4. forty-six and nine hundred fifty-five thousandths; forty-six point nine five five

5. fifty-two and one thousandths; fifty-two point zero zero one

Extension Activity

Ask students how many different ways they could write a decimal number using words.

1. The Track Banquet

A track coach has to do more than just prepare the team for track meets. Another responsibility may be to speak at a banquet at the end of the season. Imagine you are a track coach preparing a speech for the upcoming banquet. In the speech, you want to mention all the school records set this year. This means you'll have to read decimal numbers out loud. Follow the directions below to make sure you are prepared for your speech.

For each school track record listed below, use words to write it out in two different ways.

Example: In the 100-meter race, Sam Speedy set a new record of 10.24 seconds.
- "ten and twenty-four hundredths of a second"
- "ten point two four seconds"

1. In the 100-meter race, Hideo Ikeda set a new record of 10.91 seconds, and Nicole Devers set a new record of 12.01 seconds.

2. In the 400-meter race, Shawnel Johnson set a new record of 44.312 seconds, and Gwen Dawson set a new record of 51.83 seconds.

3. In the 110-meter hurdles, Alex Santiago set a new record of 14.05 seconds.

4. The girls set a new school record in the 400-meter relay of 46.955 seconds.

5. Michael O'Brien set a new school record in the 400-meter hurdles of 52.001 seconds.

Reading and Writing Decimals and Percents

teacher's page

2. Cover Design

Context

entertainment

Topic

reading and writing decimals and percents

Overview

In this activity, students listen in on a conversation between a video game designer and a graphic artist. Students are to identify the mathematics used in the conversation.

Objectives

Students will be able to:

- change decimal numbers from words to numbers
- change fractions to equivalent decimals

Materials

- one copy of the Activity 2 handout for each student

Teaching Notes

- This activity works best for individual students.
- The passage presents a hypothetical conversation between a video game designer, Malik, and a graphic artist, Anne. They are discussing the cover design for a computer software game box.
- The last two sentences refer to fractions, so students will have to convert the fractions to decimals.

Answers

1. 0.1
2. 0.3
3. 0.25
4. 10%
5. 15%
6. 0.95
7. 1.5
8. 1.75

Variation

Instead of having students work individually on this activity, have them work in pairs by reading the passages out loud to each other.

© 2007 Walch Publishing   *Real-Life Math: Decimals and Percents*

Name _____   Date _____

2. Cover Design

Designing video games requires many skills. One important skill is being able to communicate ideas effectively using mathematics. Read the conversation below between Malik, a video game designer, and Anne, a graphic artist. As you read, circle each instance where a decimal number or percent is written. For each decimal or percent that you circled, write it below using numbers.

> Malik: "The sword is too small. It needs to be about a tenth of an inch wider, and maybe three-tenths of an inch longer."
>
> Anne: "Okay, but that will push the shield to the left about twenty-five hundredths of an inch. Will that be all right?"
>
> Malik: "That's okay, but will the helmet be too small? Maybe it has to be enlarged, about ten or fifteen percent, I suppose."
>
> Anne: "We can do that, but now I'm wondering if the emblem on the shield is just a bit too large. I think I need to reduce it by a little less than one centimeter; about ninety-five hundredths of a centimeter should do it."
>
> Malik: "That should work. The width of the shield is about an inch and one-half."
>
> Anne: "Actually, the shield is closer to an inch and three-quarters."

1. _____   5. _____
2. _____   6. _____
3. _____   7. _____
4. _____   8. _____

In the space below or on a separate sheet of paper, explain the different ways that Malik and Anne referred to decimal numbers in their conversation. Also explain which method of referring to decimal numbers makes the most sense to you.

Reading and Writing Decimals and Percents

teacher's page

3. Tax Time

Context

money

Topic

reading and writing decimals and percents

Overview

In this activity, students role-play as a manager and tax accountant discussing a tax return. In the conversation, decimal numbers and percents play an important part.

Objectives

Students will be able to:

- read out loud and write decimal numbers using words
- recognize the relationship between decimal numbers and money

Materials

- one copy of the Activity 3 handout for each student

Teaching Notes

- Students should work in pairs on this activity. Have them choose a character and read those lines out loud to each other.

- Either have students record each number as it is read, or students can go back and look at the text and then write down the numbers.

- Students can work together to develop lines for the second part of the activity.

Answers

1. 645.22
2. 640.15
3. 640.15
4. 59,410.23
5. 53,219.33
6. 921.68

Extension Activity

Ask students to consider other occupations that rely on numbers, particularly decimal numbers.

© 2007 Walch Publishing

Real-Life Math: Decimals and Percents

Name _____   Date _____

3. Tax Time

Accountants often have to read, write, and speak about decimal numbers and percents. The conversation presented below could be a typical conversation between a tax accountant and her manager. In the conversation, look for each instance where a decimal number or percent is used. Each time, write it in the space below using numbers.

Manager:	"I have reviewed the tax return you gave me, and I want to discuss some questions that I have about it. First, for wages and salaries on line 7, I came up with six hundred forty-five and twenty-two hundredths, not six hundred forty and fifteen hundredths."
Tax accountant:	"I based the six hundred forty dollars and fifteen cents on the W-2."
Manager:	"Well, okay, I'll take another look. On line 22, total income should be adjusted to read *fifty-nine thousand four hundred ten point two three.*"
Tax accountant:	"That's an easy change to make."
Manager:	"I have two more questions. On line 32, shouldn't the adjusted gross income be fifty-three thousand two hundred nineteen and thirty-three hundredths, and on line 63 the amount to be refunded equals nine two one point six eight?"
Tax accountant:	"All right, I'll have the changes on your desk by the end of the day."

1. _____   4. _____

2. _____   5. _____

3. _____   6. _____

In the space below or on a separate sheet of paper, add four more lines that contain decimal numbers to this conversation. Then make up four lines of an original conversation that uses decimal numbers.

Reading and Writing Decimals and Percents

teacher's page

4. Finding Your Way

Context

travel/transportation

Topic

reading and writing decimals and percents

Overview

In this activity, students explore how decimal numbers are used when giving directions to a location.

Objectives

Students will be able to:

- read and write decimal numbers using words and numbers
- convert fractions and decimals
- accurately estimate distances using decimals

Materials

- one copy of the Activity 4 handout for each student

Teaching Notes

- This activity works best for individual students.
- Ask students if they have ever been given directions and didn't understand them, or if they have ever gotten lost because they were confused about the way decimals and fractions were used in the directions.

Answers

1. 0.3
2. 1.5
3. 3.2
4. 0.75
5. zero point three, or point three
6. three and two-tenths

Extension Activity

Have students write a statement that uses decimal and fraction words to give directions to a location such as a school or a restaurant.

4. Finding Your Way

Giving someone directions often requires using decimal numbers. For instance, you might tell someone, "Turn left at the light, go five-tenths of a mile, and my house will be on the right." Of course, if the person receiving the directions doesn't understand what five-tenths of a mile equals, he or she probably won't find your house. As you work through the problems below, you will change decimals and fractions that are written as words to numbers.

In each statement, pick out the decimal number that is written using words and write it as a number. Then answer the questions.

1. "Go to Anderson Lane, turn left, and go another three-tenths of a mile."

2. "After you cross the railroad tracks, turn left. The restaurant will be about one point five miles down the street."

3. "Pay close attention to your odometer. The turnoff is exactly three point two miles after the intersection."

4. "After the stop sign, you go another three-quarters of a mile and then turn left on Broadway."

5. How else could you write *three-tenths* using words?

6. How else could you write *three point two miles* using words?

5. Swish

Context

sports

Topic

decimals and percent equivalents

Overview

In this activity, students learn about decimal and percent equivalents by shooting free throws, gathering data, and making calculations.

Objectives

Students will be able to:

- change decimals to equivalent percents
- change percents to equivalent decimals
- change fractions to equivalent decimals

Materials

- one copy of the Activity 5 handout for each student
- sports sections from newspapers
- basketballs (one for every three or four students)
- access to basketball court

Teaching Notes

- For most of the activity, students can work in pairs or small groups. Prior to shooting the free throws, assign the students specific roles such as shooter, rebounder, and record keeper, and have them rotate. When students are ready to make their calculations, have them work on their own.

- Before using this activity, arrange to use a basketball court and balls.

- Make sure students understand that this activity is not a competition to see who can make the most free throws, and that any score will work. If a student is having a difficult time, arbitrarily choose a score by rolling dice or drawing a number.

- Box scores from basketball games usually list the number of free throw attempts and the number of free throws made for each team. Some box scores also list the decimal equivalent. Students may need help finding the required information in the box scores.

- For this activity, any box score from any basketball game will work. You might choose to use professional, college, or school game results.

(continued)

Decimals and Percent Equivalents

teacher's page

5. Swish

Answers

Answers will vary depending on individual student results and the basketball game selected in the newspaper.

Extension Activities

- Have students collect the free throw data for the entire class and calculate a class score as a decimal and as an equivalent percent.

- When students are attending a basketball game, have them keep track of their team's free throws attempted and the number made. Then have them use that information to calculate a free throw percentage for the team.

Name _____ Date _____

5. Swish

Free Throws

1. Conduct an experiment. Go to the gym and, standing at the free throw line, shoot the basketball 10 times. On a sheet of paper, record how many times you score a basket. Record your results below expressed as a fraction.

2. Change the fraction you came up with to an equivalent decimal. For example, if you scored 7 baskets, or 7 out of 10, you would divide 7 by 10, which would equal 0.7.

3. Now change your decimal score to an equivalent percent. To do this, multiply the decimal score by 100%. For instance, if your decimal score was 0.7, then 0.7 × 100% would equal 70%. In other words, you made a basket 70 percent of the time.

4. Combine the results (number of baskets scored) from three or four of your classmates. Calculate a decimal and an equivalent percent score for that group of students.

    decimal score: _____   percent score: _____

Sports Page

5. In the sports section of your newspaper, look up the box score of a basketball game. For each team, record the total number of free throw attempts and the total number of free throws the team made. List the results as a decimal. Then calculate the equivalent percent.

    Team 1 name: _____

    number of free throw attempts: _____

    number of free throws made: _____

    decimal amount: _____

    percent equivalent: _____

(continued)

5. Swish

Team 2 name: _____

number of free throw attempts: _____

number of free throws made: _____

decimal amount: _____

percent equivalent: _____

6. How many more free throws would each team from question 5 have to make in order to increase their team's free throw percentage to 90%? Explain your reasoning.

Decimals and Percent Equivalents

teacher's page

6. Doing Your Part

Context

money

Topic

decimals and percent equivalents

Overview

In this activity, students learn about decimals and percent equivalents by exploring federal income tax rates and voter turnout rates in recent presidential elections.

Objectives

Students will be able to:

- change percents to equivalent decimals
- change decimals to equivalent percents

Materials

- one copy of the Activity 6 handout for each student

Teaching Notes

- This activity works best for individual students.
- Begin by asking students to estimate what percentage of their income or their parents' income goes toward federal taxes. Then have them estimate the percentage of eligible voters who voted in the recent presidential election.

Answers

1. 10% = 0.10   28% = 0.28
   15% = 0.15   33% = 0.33
   25% = 0.25   35% = 0.35
2. 1944, .56 = 56%
   1948, .511 = 51.1%
   1952, .616 = 61.6%
   1956, .593 = 59.3%
   1960, .628 = 62.8%
   1964, .619 = 61.9%
   1968, .609 = 60.9%
   1972, .5521 = 55.21%
   1976, .535 = 53.5%
   1980, .54 = 54%
   1984, .531 = 53.1%
   1988, .502 = 50.2%
   1992, .559 = 55.9%
   1996, .49 = 49%
   2000, .513 = 51.3%
   2004, .553 = 55.3%

Extension Activities

- Have students research occupations and salary levels to determine which tax rate would apply for that particular job.
- Have students research state income tax rates for your state and answer similar questions based on those rates.

6. Doing Your Part

1. The table below lists the 2005 individual income tax rate percentages and taxable income level for people who are single. In the last column, write the decimal equivalent for each tax rate percentage.

2005 Individual Income Tax Rate for Singles

Taxable income level	Tax rate (percent)	Tax rate (decimal)
$0–$7300	10%	
$7301–$29,700	15%	
$29,701–$71,950	25%	
$71,951–$150,150	28%	
$150,151–$326,450	33%	
$326,451 and up	35%	

Source: http://www.irs.gov/

2. The table below lists the voter turnout in presidential elections from 1944 to 2004. For each election year, voter participation is listed as a decimal amount of the voting-age population. Change the voter participation decimal amounts to equivalent percentages and write them in the table.

Voter Turnout in Presidential Elections, 1944–2004

Year	Voter participation (decimal amount of voting-age population)	Voter participation (% amount of voting-age population)	Year	Voter participation (decimal amount of voting-age population)	Voter participation (% amount of voting-age population)
1944	.56		1976	.535	
1948	.511		1980	.54	
1952	.616		1984	.531	
1956	.593		1988	.502	
1960	.628		1992	.559	
1964	.619		1996	.49	
1968	.609		2000	.513	
1972	.5521		2004	.553	

Source: http://www.infoplease.com/

© 2007 Walch Publishing — Real-Life Math: Decimals and Percents

Decimals and Percent Equivalents

teacher's page

7. Who's Watching What?

Context
entertainment

Topic
decimals and percent equivalents

Overview
In this activity, students assume the role of a television marketing manager who must change decimals to equivalent percents in order to complete a report.

Objective
Students will be able to:
- change decimals to equivalent percents

Materials
- one copy of the Activity 7 handout for each student

Teaching Notes
- This activity is best for individual students.
- A way to start this activity is to have students estimate the percentage of homes in the United States that are watching the most popular broadcast television programs.
- Explain to students that according to Nielsen Media Research, there are more than 110 million television households in the United States. A single ratings point is the equivalent of 1% of those households, or 1,100,000 homes.

Answers
1. 18.2%
2. 15.8%
3. 14.3%
4. 13.8%
5. 12.5%
6. 12.3%
7. 12.3%
8. 12.0%
9. 11.1%
10. 11.0%

Extension Activity
Have students calculate (using 110 million homes) the total number of households tuned in to a particular show.

© 2007 Walch Publishing — Real-Life Math: Decimals and Percents

Name _____ Date _____

7. Who's Watching What?

Suppose you are a marketing manager for a major television network. Your boss wants you to find out what percentage of homes are watching the 10 most popular broadcast television shows for the week of 11/07/05 to 11/13/05. Unfortunately, as you research that information, you can only find the week's ratings listed as decimal amounts. To complete the report for your boss, change the decimal ratings into equivalent percentages. List them in the last column of the table.

List of Week's Broadcast TV Ratings from 11/07/05–11/13/05

Rank	Program	Rating (decimal)	Rating (percentage)
1	CSI	.182	
2	Desperate Housewives	.158	
3	NFL Monday Night Football	.143	
4	Without a Trace	.138	
5	Grey's Anatomy	.125	
6	CSI: Miami	.123	
6	CSI: NY	.123	
8	Lost	.120	
9	Cold Case	.111	
10	NCIS	.110	
10	Survivor: Guatemala	.110	

Source: www.nielsenmedia.com/ratings/

Decimals and Percent Equivalents

teacher's page

8. Metric Measures

Context

travel/transportation

Topic

decimals and percent equivalents

Overview

In this activity, students practice converting common decimals to percent equivalents and common percents to decimal equivalents using the metric system.

Objectives

Students will be able to:

- change percents to equivalent decimals
- change decimals to equivalent percents

Materials

- one copy of the Activity 8 handout for each student

Teaching Notes

- This activity works best for individual students.
- Students should be familiar with the metric system or have access to a metric system guide.

Answers

1. 40 centimeters are equal in length to 40% of 1 meter.
2. 500 meters are equal in length to 0.5 kilometer.
3. 0.25 of 1 kilometer is equal in length to 25 decameters.
4. 75 centimeters are equal in length to 75% of 1 meter.
5. $0.\overline{3}$ of 1 centimeter is equal in length to $3\frac{1}{3}$ millimeters.
6. $66\frac{2}{3}\%$ of 1 meter is equal in length to $0.\overline{6}$ meter.

Extension Activities

- In addition to rewriting the statements as decimals or percents, have students rewrite the statements using fractions.
- Have students make up their own problems and have classmates solve them.
- Set up similar problems using a different measurement system.

8. Metric Measures

One of the most practical skills you can learn is converting—without paper and pencil—common decimals to equivalent percents, and common percents to equivalent decimals. It takes practice to become proficient at these skills. One way to practice is to examine relationships within the metric system.

Rewrite each statement below, changing the underlined percent or decimal to an equivalent decimal or percent.

Example: 1 decimeter is equal in length to 0.1 meter.
Rewrite the sentence using a percent: 1 decimeter is equal in length to 10% of 1 meter.

1. 40 centimeters are equal in length to 0.4 meter.

2. 500 meters are equal in length to 50% of 1 kilometer.

3. 25% of the length of 1 kilometer is equal in length to 25 decameters.

4. 75 centimeters are equal in length to 0.75 meter.

5. $33\frac{1}{3}$% of 1 centimeter is equal in length to $3\frac{1}{3}$ millimeters.

6. $66\frac{2}{3}$ centimeters are equal in length to $.666\overline{6}$ meter.

Comparing Decimal Numbers

teacher's page

9. The Batting Order

Context

sports

Topic

comparing decimal numbers

Overview

In this activity, students become high-school baseball coaches who must rely on players' statistics in order to make out a batting lineup.

Objective

Students will be able to:

- compare and order decimals

Materials

- one copy of the Activity 9 handout for each student
- nine small pieces of paper for each student, pair, or group

Teaching Notes

- Students can work individually, in pairs, or in small groups.
- Statistics are listed for only nine players in order to reduce the possible lineup choices. Students should assume that the players' fielding positions have already been determined.
- Use the small pieces of paper to make up cards with each player's statistics on it. Once the cards are made, select the highest averages in each category and place those players in their respective spots. Then use the guidelines in Table 2 to complete the lineup.

Answers

More than one possible lineup meets the given criteria. However, the first, third, and fourth hitters must always match the sample lineup.

Sample lineup:

1. Baker
2. Sullivan
3. Barone
4. Wong
5. O'Connor
6. Martinez
7. Gamage
8. Hardy
9. Voornas

Extension Activities

- Have students research other sports where generally accepted mathematical criteria determine a player's position.
- During baseball season, have your students construct a table similar to Table 1. Have them compare an actual team's statistics with the criteria found in Table 2.

© 2007 Walch Publishing

Real-Life Math: Decimals and Percents

9. The Batting Order

Imagine you have just been hired as the new high-school baseball coach. Unfortunately, it is the middle of the season, you have a game tomorrow, and there is no one at the school who can help you create the batting lineup. That means you will have to rely entirely on the players' statistics. Based on your coaching experience, you know that a player's batting average, on-base average, and slugging average are good indicators of where a player should bat in the lineup. Listed below are the current season's statistics for each player. Table 2 has the criteria to be used to place the players in the lineup. Make up cards with each player's statistics and arrange them in a batting order that is most closely aligned to the given criteria. When you are finished, create a lineup card and turn it in to your teacher.

Player Season Statistics (Table 1)

Player's name	Batting average	On-base average	Slugging average
Martinez	.312	.337	.412
Baker	.417	.479	.361
Gamage	.299	.333	.353
Wong	.340	.357	.612
Barone	.431	.463	.584
O'Connor	.325	.362	.574
Hardy	.275	.343	.353
Voornas	.250	.307	.313
Sullivan	.425	.414	.399

Batting Order Criteria (Table 2)

Place in batting order	Batting average	On-base average	Slugging average
1	high	highest	low–medium
2	high	medium–high	medium
3	highest	high	high
4	medium	medium	highest
5	medium	medium	high
6	medium	medium	medium
7	low–medium	low–medium	low–medium
8	low–medium	low–medium	low–medium
9	lowest	lowest	lowest

Comparing Decimal Numbers

teacher's page

10. Fast-Food Wars

Context

entertainment

Topic

comparing decimal numbers

Overview

In this activity, students examine percent of market share for the top 10 fast-food restaurants.

Objectives

Students will be able to:

- compare and order percents
- construct a graph

Materials

- one copy of the Activity 10 handout for each student

Teaching Notes

- This activity works best for individual students.
- Some students may need further explanation about the concept of market share.

Answers

1. Market share order:

   1. McDonald's
   2. Burger King
   3. Pizza Hut
   4. Taco Bell
   5. Wendy's
   6. KFC
   7. Hardee's
   8. Subway
   9. Dairy Queen
   10. Domino's

2. Graphs will vary. Sample graph:

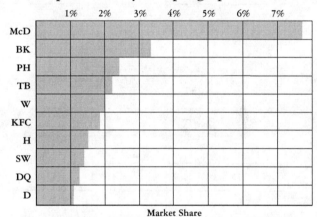

3. Answers will vary. Sample answer:

   Taco Bell has greater market share than six other fast-food restaurants but less market share than three other restaurants.

Extension Activity

Have students write the Taco Bell manager's report on Taco Bell's standing with regard to its competitors.

10. Fast-Food Wars

As a newly hired manager for Taco Bell, part of your job is to keep current with how your business is doing in relation to other fast-food restaurants. The latest market share figures for the top 10 fast-food restaurants have just been released. Follow the steps below to create a presentation for the district manager showing how your restaurant is doing compared to the competition.

1. Organize the market share information (Table 1) in order from greatest market share to least market share. List the results in Table 2. Place the highest-ranking restaurant at the top, and continue the list in descending order.

Fast-Food Restaurant Market Share Data (Table 1)

Name of restaurant	Market share	Name of restaurant	Market share
Dairy Queen	1.2%	Hardee's	1.5%
McDonald's	7.8%	Pizza Hut	2.5%
Wendy's	2.0%	KFC	1.8%
Subway	1.3%	Domino's	1.1%
Burger King	3.4%	Taco Bell	2.2%

Top 10 Fast-Food Restaurants by Percent of Share of Market (Table 2)

Rank	Name of restaurant	Market share (percent)

(continued)

10. Fast-Food Wars

2. After ranking the restaurants by market share, you decide to use a graph in your presentation. Use the space below to create the graph. List the names of the restaurants on the vertical axis. Show the percents of market share on the horizontal axis.

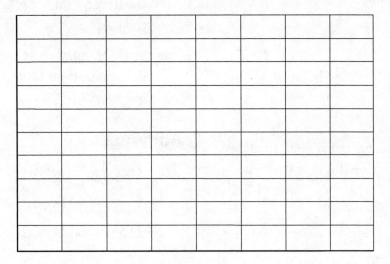

3. Describe Taco Bell's percent of market share in relation to the other fast-food restaurants.

Comparing Decimal Numbers

teacher's page

11. Which Costs Less: UPS or the U.S. Postal Service?

Context

travel/transportation

Topic

comparing decimal numbers

Overview

In this activity, students compare decimal numbers to determine the lowest shipping costs.

Objectives

Students will be able to:

- determine order of decimal numbers
- add decimal numbers
- construct a graph

Materials

- one copy of the Activity 11 handout for each student

Teaching Notes

- This activity works best for individual students.
- Introduce the activity by having students list the factors that determine how much it costs to ship a package.
- The shipping costs listed in Table 1 are rates for shipping a package between California and Texas, about 1000 miles.
- If students are struggling with questions 5 and 6, have them construct a table to help them compare the costs.
- Students may need help organizing the information on their graphs in question 7.

Answers

1. U.S. Postal Service, $65.05
2. U.S. Postal Service, $65.90
3. U.S. Postal Service, $24.48
4. UPS, $17.60
5. Ship both via the U.S. Postal Service for $124.12.
6. Ship both via the U.S. Postal Service; total cost is $145.73.
7. Graphs will vary.

Extension Activity

Have students investigate shipping prices for other shipping companies, such as DHL.

Data Resources

http://postcalc.usps.gov

www.ups.com

© 2007 Walch Publishing

Real-Life Math: Decimals and Percents

11. Which Costs Less: UPS or the U.S. Postal Service?

Imagine that you are packing your suitcase to spend your summer with your grandparents. As you pack, it becomes obvious that you will not have enough space for all your stuff. You decide to ship some items. Of course, you want to ship the packages for the least amount of money. Examine the shipping costs listed below in Table 1 for United Parcel Service (UPS) and the United States Postal Service (U.S. Postal Service). Then answer the questions that follow.

Shipping Costs for UPS and U.S. Postal Service (Table 1)

Package size (lbs.)	UPS			U.S. Postal Service		
	5-day delivery	2-day delivery	Overnight delivery	5-day delivery	2-day delivery	Overnight delivery
25	$17.60	$71.22	$121.34	$17.83	$28.05	$65.05
45	$27.81	$99.18	$166.29	$24.48	$46.55	$97.40
65	$34.72	$134.91	$212.98	$26.72	$65.90	$130.85

Source: United States Postal Service, United Parcel Service

1. If you have a 25-pound package that has to be sent overnight, which shipping method is less expensive, and how much does it cost?

2. If you have a 65-pound package that must arrive in 2 days, which shipping method is less expensive, and how much does it cost?

3. You have a 45-pound package that can arrive in 5 days. Which shipping method is less expensive, and how much does it cost?

(continued)

11. Which Costs Less: UPS or the U.S. Postal Service?

4. You decide that your 25-pound package can arrive in 5 days. Which shipping method is less expensive, and how much does it cost?

5. You have a 45-pound package that has to be sent overnight and a 65-pound package that can arrive in 5 days. Which shipping method or combination of methods should you choose so that each package is shipped for the lowest cost?

6. You have a 25-pound package that has to be sent overnight and a 45-pound package that must arrive in 2 days. Which shipping method or combination of methods should you choose so that each package is shipped for the lowest cost?

7. Construct a graph in the space below that shows the costs for each delivery schedule and weight. Label the weights and the delivery schedules for both organizations on the horizontal axis. Label the vertical axis "Cost."

Comparing Decimal Numbers

teacher's page

12. Dividing Up Lunch

Context

money

Topic

comparing decimal numbers

Overview

In this activity, students have to estimate meal costs to determine how much each person owes for lunch.

Objectives

Students will be able to:

- choose appropriate strategy for determining lunch costs
- round decimal numbers in order to estimate lunch costs
- add and subtract decimal numbers

Materials

- one copy of the Activity 12 handout for each student

Teaching Notes

- Students can work individually, in pairs, or in small groups.

- Depending on your students' background, they may not have experience with this type of situation; some students may need more context to understand it.

- Some students will want to add up exact costs. Help those students estimate to simplify the task.

- This activity can lead to an interesting discussion about figuring the appropriate place value for rounding a number. For instance, should the side salad be rounded to $2.75 or $2.50? What are some of the factors that might determine this?

Answers

Answers will vary depending on how students determine each person's share.

One possible set of answers:
You: $11.25
Joe: $6.25
Tyson: $12.00
Jill: $8.00
Keiko: $7.00

Extension Activities

- Have students add an additional 15–20% onto each person's share to cover the tip.

- Next time students go to a restaurant, ask them to estimate what each person will owe.

12. Dividing Up Lunch

You and four friends are having lunch together at a local diner. The check arrives and it's time to figure out who owes what. Since you are the math expert in the group, everyone turns to you to figure out how much they should contribute. Each person's meal and its cost is listed below. Review what each person ate and then answer the questions that follow.

You	Joe	Tyson	Jill	Keiko
hamburger $4.85	turkey sandwich $4.25	cheeseburger $5.05	chicken salad $6.95	pastrami sandwich $4.75
side salad $2.65	chips $0.90	side salad $2.65	tea $0.95	fries $1.25
tea $0.95	tea $0.95	large soda $1.15		small soda $0.85
cake $2.75		cheesecake $2.95		

1. Briefly explain how you will determine each person's share of the check.

2. Once you have decided on a method, determine each person's share. List the amounts below.

3. Explain why there is a difference between the amount of money you collected from your friends and the check's actual total.

4. Compare how much you are asking each person to pay with a classmate's results. Discuss the differences.

© 2007 Walch Publishing   Real-Life Math: Decimals and Percents

Basic Operations with Decimals

teacher's page

13. The Building Challenge

Context

money

Topic

basic operations with decimals

Overview

In this activity, students work as a team of architects who are preparing a proposal to build an office building. As part of the activity, students will construct a model of the building and calculate costs.

Objectives

Students will be able to:

- work cooperatively as a team
- add, subtract, multiply, and divide decimal numbers

Materials

- one copy of the Activity 13 handout for each student
- plastic straws, approximately 40 straws per group
- masking tape, one or two rolls per class
- paper clips, approximately 20 per group (optional)
- tape measure (must have metric measures) or metersticks, one per group

Teaching Notes

- Students should work in small groups for this activity.
- During the activity, you will act as the Proposal Chairperson and review designs prior to any construction. Also, you will have to verify the total cost of each structure and select a winning proposal that most closely meets the criteria.
- The minimum building cost will ensure that the students' designs will not be too simplistic.
- Before the activity, review with your students how a frequency table is used.
- You may choose to set a time limit of 30 to 45 minutes for the construction phase.
- Cardboard boxes or poster board can be used as bases for the structures.

Answers

Answers will vary depending on the amount of building materials used.

Extension Activity

Have students conduct research about architecture and building designs.

Name _____   Date _____

13. The Building Challenge

Imagine you are part of a team of architects preparing a proposal for designing an office building. In order for your team's proposal to be accepted, it has to meet the building and cost specifications listed below, and cost less than the other proposals submitted. The first step in preparing the proposal is coming up with a design and drawing it on paper. Then it will be necessary to build a model. Remember, the winning proposal will be a building that is kept within the height range and is as close to the minimum cost as possible.

> **Building specifications:**
>
> - Height must be between 250 and 275 centimeters.
> - Building cannot be supported by other structures (walls, ceilings, and so forth). It must stand on its own.
>
> **Building material costs:**
>
> straws: $1.13 each
>
> masking tape: $0.31 per centimeter
>
> paper clips: $0.25 each
>
> **Minimum building cost:**
>
> Building must cost at least $35.00 to build.

Step 1

Decide on a design with your group. Make a sketch of the building on a separate sheet of paper. Next, estimate material costs to make sure that you are at or above the minimum building cost. When you have completed step 1, submit the drawing to the Proposal Chairperson for review.

Step 2

Construct your building. Use the frequency table below to maintain an accurate record of material costs during construction.

Item	Tally	Frequency
straws		
paper clips		
centimeters of tape		

(continued)

Name _____ Date _____

13. The Building Challenge

Step 3

Calculate building material costs.

Number of item	Item description	Cost
	straws × $1.13 =	
	cm of tape × $0.31 =	
	paper clips × $0.25 =	
	Total cost =	

Step 4

Find the difference between the cost of the building materials you used and the minimum building cost.

Step 5

List the height of your model in centimeters.

Step 6

Submit your proposal for consideration to the Proposal Chairperson.

Basic Operations with Decimals

teacher's page

14. Money Exchange

Context

travel/transportation

Topic

basic operations with decimals

Overview

In this activity, students learn about converting between various currencies to determine if they have paid appropriate prices for items.

Objective

Students will be able to:

- perform basic calculations with decimal numbers

Materials

- one copy of the Activity 14 handout for each student

Teaching Notes

- This activity works best for individual students.
- Students may require an initial introduction to currency exchange.

Answers

1. Correct; the converted price is $139.54.
2. Incorrect; you were charged the equivalent of $67.72, almost twice the predicted cost.
3. Incorrect; you were charged $20.28 for lunch.
4. Incorrect; you paid $14.94.
5. Correct; the meal came to $120.38.
6. Correct; your purchases cost $24.30.
7. Correct; your purchases cost $12.97.

Extension Activities

- Have students perform Internet research on exchange rates for other countries and currencies.
- Have students track and graph the historical fluctuation of exchange rates. You can also ask students to investigate the causes of those fluctuations.

14. Money Exchange

As part of Aiden's sales job with his new company, he has to travel all around the world, spending only a day or two in each country. With all that travel, it is difficult to keep track of where he is, how much things should cost, and what currency he should pay with. Listed below are the exchange rates of several countries Aiden has recently visited. Look over the exchange rates, then read the problems below. In each case, decide if the amount Aiden was charged matches the estimate of the price in U.S. dollars. If the estimate was correct, write **correct** and the amount he was charged in U.S. dollars. If you think the estimate was incorrect, write **incorrect** and the amount he was charged in U.S. dollars.

Currency Exchange Rates
(National currency units per 1 U.S. dollar)

Country (currency)	Britain (pound)	France (euro)	Switzerland (franc)	Russia (ruble)	China (yuan)	Japan (yen)	Mexico (peso)
Exchange rate	0.5303	0.7826	1.2329	26.778	7.975	115.23	10.98

Source: www.xe.com

1. The first night, Aiden stayed in a hotel in London, England. He was charged 74 pounds. His guide book said that the price at that hotel should have been between $125 and $145.

2. After flying to Paris, France, Aiden took a cab to see the Eiffel Tower. His cabbie charged him 53 euros. Aiden estimated that a cab from the airport to downtown (near the Eiffel Tower) should have cost about $40.

3. As a day trip, Aiden took the bus from France to Geneva, Switzerland. He estimated that his bill for a sandwich and a cup of tea at a sidewalk café in Geneva was $10. He was charged 25 Swiss francs.

(continued)

14. Money Exchange

4. After returning to Paris, Aiden flew to Moscow for a meeting. He decided to buy a shirt for his mom and chose what he thought was a $20 T-shirt. He was charged 400 rubles.

5. From Russia, Aiden traveled to China, where he hosted a nice dinner for a client. The concierge at the hotel indicated that dinner would cost about $60 per person at this particular restaurant. The bill for the two of them came to 960 yuan.

6. Aiden's flight from China stopped in Tokyo, Japan, where he bought some postcards and a book. He expected the bill to be about $24. The clerk charged him 2800 yen.

7. Aiden's flight home was diverted to Mexico City due to bad weather. While there, he grabbed a magazine and a bite to eat in the airport. Aiden estimated the bill would be less than $14. He paid 142.4 pesos.

Basic Operations with Decimals

teacher's page

15. Scorekeeper

Context
sports

Topic
basic operations with decimals

Overview
In this activity, students are acting as parents who have volunteered to be the official scorekeepers at their son's and daughter's sporting events.

Objectives
Students will be able to:
- add, subtract, multiply, and divide decimal numbers
- find averages
- round decimals

Materials
- one copy of the Activity 15 handout for each student

Teaching Notes
- This activity works best for individual students.
- Scoring for figure-skating and diving contests differs depending on the level of competition and the organizing athletic body.
- For this activity, figure-skating scores will be out of a possible 6.0, and diving scores will be out of a possible 10.0.
- Averages should be carried to the thousandth place.

Selected Answers

Figure skating

Round 1: 4.975

Round 2: 4.150

Round 3: 5.475

Overall average: 4.867

Diving

First dive: 8.467

Second dive: 9.167

Third dive: 8.367

Fourth dive: 8.133

Fifth dive: 9.533

Overall average: 8.733

Extension Activity
Have students watch an actual figure-skating contest or a diving meet to see the scoring process in action.

Name _____   Date _____

15. Scorekeeper

Having a son or a daughter who participates in sports means that you get to volunteer your time and help out at meets and games. Imagine your daughter has a figure-skating contest Saturday morning, and you have volunteered to act as official scorekeeper. That afternoon, your son will be diving at a swim meet where you also have volunteered to act as official scorekeeper. The scorekeeper's job is to record the judges' scores for each round, then calculate the average score for that round and a final overall average score.

Figure-Skating Scores

	Judge 1	Judge 2	Judge 3	Judge 4	Round avg.
Round 1	4.8	4.8	5.1	5.2	
Round 2	3.9	4.1	4.2	4.4	
Round 3	5.3	5.4	5.7	5.5	

1. Find the average score for each round and list that amount in the table.
2. Find the overall average score for all three rounds and write it below.

   overall average score: _____

Diving Scores

	Judge 1	Judge 2	Judge 3	Round avg.
First dive	8.6	8.7	8.1	
Second dive	9.1	9.0	9.4	
Third dive	8.3	8.7	8.1	
Fourth dive	7.9	8.2	8.3	
Fifth dive	9.4	9.6	9.6	

1. Find the average score for each dive and list that amount in the table.
2. Find the overall average score for all five dives and write it below.

   overall average score: _____

© 2007 Walch Publishing   Real-Life Math: Decimals and Percents

Basic Operations with Decimals

teacher's page

16. Movie Night

Context
entertainment

Topic
basic operations with decimals

Overview
In this activity, students go out for pizza and to the movies, but first they must estimate how much it is going to cost.

Objectives
Students will be able to:
- add, subtract, multiply, and divide decimal numbers
- round decimals

Materials
- one copy of the Activity 16 handout for each student

Teaching Notes
- This activity works best for individual students.
- Ask students to think about how much money they spent the last time they went out to dinner and the movies.

Answers
1. 2.3 gallons
2. $6.69
3. $2.23
4. The total bill is $38.85, and each person's share is $12.95.
5. $15.50
6. $8.65
7. $49.33
8. $60.00
9. 7 hours

Extension Activity
Have students estimate expenses prior to going out and then have them compare actual expenses to their estimations.

Name _____   Date _____

16. Movie Night

It's been a long, hard week at school. Saturday night you and two friends decide to get some pizza and go to the movies. However, before you go, you need to figure out how much money you should withdraw from the ATM. Answer the questions below to help you determine how much money to take out.

1. First, you have to borrow the car from your parents. They require you to pay for your own gas. You estimate that the trip to pick up your friends, go to the restaurant and the movie theater, and then return home will be about 55 miles. Your parents' car gets 24 miles to the gallon for this kind of drive. How many gallons of gas will you need to buy?

2. If gas costs $2.909 per gallon, how much money will gas cost for the trip?

3. You will ask your two friends to split the cost of gas three ways. How much is each person's share?

4. You and your friends will order your usual at the pizza place—two large pizzas, one cheese and one with pepperoni and mushrooms. You will also get a pitcher of soda. A large cheese pizza is $12.95 and toppings are $1 each. The pitcher of soda is $4.95. You usually leave the waiter a $6.00 tip. If you split the bill three ways, how much is each person's share?

(continued)

© 2007 Walch Publishing   *Real-Life Math: Decimals and Percents*

16. Movie Night

5. A movie ticket at the new theater with stadium seating costs $8.75. But a movie isn't a movie without popcorn for $3.50 and a soda for $3.25. How much will you spend at the movie theater?

6. After the movies, you and your friends will probably each spend about $5.00 on video games and $3.65 on ice cream at the local ice cream shop. How much will you spend after the movies?

7. Look over your anticipated expenses above. After totaling them, you decide to bring an extra $10, since your friend Blake never brings enough money. How much will you need to cover your expenses plus $10 extra for Blake?

8. The ATM at your bank only dispenses $20 bills. How much will you need to withdraw to pay for the evening?

9. You get a part-time job in order to be able to afford your social life. The job pays $7.35 per hour. How many full hours would you have to work in order to earn enough for another night out like this one?

Percent of a Number

teacher's page

17. TV Time

Context

entertainment

Topic

percent of a number

Overview

In this activity, students construct circle graphs, based on given data and data collected while watching television.

Objectives

Students will be able to:

- find percent of a number
- construct circle graphs

Materials

- one copy of the Activity 17 handout for each student
- protractors
- stopwatch/timer

Teaching Notes

- This activity works best for individual students.
- When making circle graphs, students often have difficulty visualizing that 360 degrees of a circle equals 100% of that circle. Students may see this relationship more clearly if you demonstrate changing a familiar amount, such as 25%, to degrees before students attempt it on their own.
- Emphasize that students must follow each step carefully when constructing their circles.
- Have students make all their calculations prior to drawing any sectors on their circles.
- Make sure that students draw the first sector with the protractor edge on zero degrees. Then they should draw subsequent sectors using the "top" of the sector as the zero degree line.
- Have students watch network programming when they collect their data, and have them avoid premium channels, pay-per-view channels, sporting events, and so forth.

Answers

Answers will vary according to students' data.

Extension Activity

After students construct their circle graphs by hand, use a spreadsheet program to compare how the graphs look.

Name _____   Date _____

17. TV Time

Circle graphs help you visualize the amount or share of related items to the whole in terms of percent. A whole circle represents 100%, and each individual category listed in the graph represents that particular category's percent out of the total 100%.

When you watch television, what percent of the time do you think you spend actually watching the program? Let's find out. As you watch television tonight, make a chart to keep track of commercials, station breaks and previews, and the program itself. Do this by noting the time when each program, commercial, or preview starts and stops. Do this for one hour. When you are finished watching, calculate what percent of time each category represents. Construct a circle graph showing the results. Then answer the questions that follow.

Steps for constructing a circle graph:

1. Change the percent amount to a decimal.

    Example: 24.8% × 0.248

2. Multiply the decimal amount by 360 to find the number of degrees.

    Example: 0.248 × 360 = 89.28°

3. Round the number of degrees to a whole number.

    Example: 89.28° × 89°

4. Use your protractor to draw a sector equal to the number of degrees on your graph.

Television Time in Percents

program: _____   preview: _____

commercial: _____   other: _____

station break: _____

(continued)

17. TV Time

1. On average, a teenager watches 21 hours of television per week. Based on your television data, how many hours will you spend watching commercials as a teenager?

2. How many hours of programs will you watch while you're a teenager?

Percent of a Number

teacher's page

18. Winning Percentage

Context

sports

Topic

percent of a number

Overview

In this activity, students investigate examples of how winning percentages are used in the sports section of the newspaper, and then calculate their winning percentages after playing games of five in a row.

Objectives

Students will be able to:

- change fractions to decimals
- change decimals to percents
- calculate individual and class winning percentages

Materials

- one copy of the Activity 18 handout for each student
- grid paper, 2 or 3 sheets per student
- sports section of newspaper (1 per group or pair)

Teaching Notes

- Students can work in pairs or small groups for this activity but should calculate individual and class winning percentages on their own.
- Typically, two or three games of five in a row can be played on each side of a sheet of grid paper.
- Little explanation is needed on how to play five in a row since it is essentially a more difficult version of tic-tac-toe, the difference being that you need five in a row to win.
- Encourage students to play as many different students as time permits.
- When playing time is finished, make sure students turn in their winning percentages to you so you can show them on the board or overhead.

Answers

Answers will vary depending on the number of games each student wins.

Name _____   Date _____

18. Winning Percentage

Coaches in many different sports are judged by their team's winning percentage. Winning percentage is calculated based on how often a team wins. It is expressed as either a percent or a decimal number. In other words, if Coach Castillo has a winning percentage of 75, it could be said that "Coach Castillo wins 75% of his games."

1. Look through the sports section of the newspaper and find three different examples of how winning percentages are used. Describe each example below.

    a.

    b.

    c.

2. Play as many other students as you can in games of five in a row. Use the table below to keep track of your wins and losses. When the time for playing the games has expired, calculate your winning percentage.

    Example: If you played 8 games and won 5 of them, then you would divide 8 into 5, then change the answer, 0.625, from a decimal to a percent, 62.5%. So your winning percentage would be 62.5. In other words, you won 62.5% of your games.

    wins/games = 5/8 = 0.625 = 62.5%

wins	
losses	

    your winning percentage: _____

3. Calculate the winning percentage for the class as a whole.

    class winning percentage: _____

Percent of a Number

teacher's page

19. Planning a Trip

Context

travel/transportation

Topic

percent of a number

Overview

In this activity, students use percentage rates to help them decide where to vacation.

Objectives

Students will be able to:

- change decimals to percents
- find percent of a number

Materials

- one copy of the Activity 19 handout for each student

Teaching Notes

- This activity works best for individual students.

- Explain to students that the weather data was available either as a percent or as number of days for that particular type of weather.

- The number of cloudy days is extraneous information.

- For question 1, students will find the number of clear days based on 365 days, or 1 year.

- For question 2, students first must total the number of clear and partly cloudy days to figure out which yields the highest percent.

- For question 3, students have to change percent of precipitation days to number of days.

Answers

1. New Orleans, 109 days
2. Honolulu, 72.33%
3. Honolulu, 72.33%, 102 days of precipitation
4. Honolulu

Extension Activity

Have students monitor the selected city's weather for a period of time and compare that data with the data found in the activity.

19. Planning a Trip

Imagine you are planning a well-deserved vacation. You have narrowed your choice of destinations to six cities. In making the final decision about which city to visit, you will use the following criteria: You want to vacation in a city with the highest percent of clear and partly cloudy days and the lowest number of days with precipitation. You have compiled weather data (Table 1) for the cities. Now you are ready to figure out which city meets your criteria. Unfortunately, some of your data show number of days, and some show a percent. To compare data, you will need to convert the figures.

Weather Data for Selected Cities

City	Clear days per year	Partly cloudy days per year	Cloudy days per year	Precipitation days per year
Atlantic City	26.30%	108	161	30.80%
Daytona Beach	26.58%	135	134	31.23%
Honolulu	24.66%	174	101	27.95%
Miami	20.82%	172	117	35.34%
New Orleans	29.86%	120	136	30.96%
Seattle	15.61%	79	229	43.84%

Use the information in the table to answer the following questions.

1. In which city would you find the highest percent of clear days? List the name of the city and the number of clear days.

2. In which city would you find the highest percent of clear and partly cloudy days? List the name of the city and the percent of clear and partly cloudy days.

3. In which city would you find the highest percent of clear and partly cloudy days and the lowest number of days with precipitation? List the name of the city, the percent rate, and the number of days with precipitation.

4. What city most closely meets your weather criteria?

Percent of a Number

teacher's page

20. Deductions

Context
money

Topic
percent of a number

Overview
In this activity, students examine deductions from gross pay to determine what percent of pay is eaten up by deductions.

Objectives
Students will be able to:
- change decimals to percents
- find percent of a number

Materials
- one copy of the Activity 20 handout for each student
- pay stubs, either actual stubs with names and social security numbers obliterated, or facsimile pay stubs created for this activity (They should show the categories of deductions listed in the answer to question 1.)

Teaching Notes
- Students can work individually, in pairs, or in small groups.
- This activity works best using actual pay stubs. However, obtaining pay stubs might be difficult. Students who work part-time jobs are a good source of pay stubs.
- When you use pay stubs collected from others, be especially careful to cover any identifying information.
- After you hand out copies of the pay stubs, ask students to look them over carefully, paying particular attention to the different types of deductions.
- Some students may need additional information about deductions before this activity.

Answers
1. Students should list the following categories of deductions: Social Security tax, Medicare tax (some organizations combine Social Security and Medicare into one category called FICA), federal income tax, and state income tax. They may also list medical, dental, and retirement deductions.

2. Social Security is taxed at 6.2%; Medicare is taxed at 1.45%; federal and state income tax rates vary; likewise, medical, dental, and retirement rates will vary.

3–5. Answers will vary.

Extension Activity
Students can extend the last problem by figuring out how many days or months of the year they work to pay for deductions.

20. Deductions

Look over your pay stub, paying particular attention to the different types of deductions. The money taken out of your pay is called a **deduction,** and those deductions are the difference between the amount you are paid (gross pay) and the amount you actually end up with (net or take-home pay). But where do all those deductions come from, and what percent of your pay do you never see? Good question. Work through the activity below to find out.

Use the pay stubs to answer the following.

1. List all the different types of deductions that you find.

2. For each different type of deduction, find the percentage rate used to calculate the amount of the deduction. Fill in the table on the next page with your information.

   Example: If your gross pay is $2912.12, and the amount taken out of your pay for retirement is $186.37, then you would set up a percent equation to solve for the percent rate.

   $186.37 is what percent of $2912.12?

   Let n = percent rate.

   Write "is" as an equal sign, and write "of" as a multiplication symbol.

   $186.37 = n \times 2912.12$

   $$n = \frac{186.37}{2912.12} = 0.06399$$

   Change 0.06399 to a percent, 6.4%.

   The deduction rate for retirement is 6.4% of your gross pay.

(continued)

Name _____   Date _____

20. Deductions

Types of Deductions and Percent Rate

Type of deduction			
Percent rate			
Type of deduction			
Percent rate			

3. Compare the types of deductions and the rates you calculated with those of another student. Describe the similarities and differences that you notice.

4. If you totaled up all the deductions, what percent of your pay is deducted?

5. In an 8-hour day, how many hours do you spend working to pay for deductions?

Percent of Increase/Decrease

teacher's page

21. How Much Is School Worth?

Context
money

Topic
percent of increase/decrease

Overview
In this activity, students calculate the worth of an education in dollar amounts.

Objective
Students will be able to:
- find percent of increase or decrease

Teaching Notes
- This activity works best for individual students.
- Point out that the salaries listed are averages, and individuals may earn more or less than what is in the table.

Answers
1. Students should note that salaries are on the rise.
2. 52.9% increase
3. 75.3% increase
4. 10.0% increase
5. a. $756,000
   b. $1,036,000
   c. $1,816,000
6. $780,000 or a 75.3% increase
7. 45.2% increase overall, which is 2.3% per year
8. Answers will vary depending on year of graduation.

Extension Activities
- Have students investigate other relationships within the table.
- Research college costs and compare them to annual and lifetime earnings.

© 2007 Walch Publishing

Real-Life Math: Decimals and Percents

21. How Much Is School Worth?

Did you know that the more school or education you receive, the more money you will probably make in your lifetime? It's true! But how much more money? Complete the activity below to answer that question.

The chart lists annual earnings (salaries) based on a person's education level. Look over the chart and then answer the questions that follow.

Annual Mean Earnings by Educational Attainment, 1985–2005

Year	Not a high-school graduate	High-school graduate	Some college	Bachelor's degree	Advanced degree
1985	$10,276	$14,457	$16,349	$24,877	$32,909
1995	$14,013	$21,431	$23,862	$36,980	$56,667
2005	$18,900	$25,900	$28,500	$45,400	$99,300

Source: U.S. Census Bureau

1. Describe the trend in salaries from 1985 to 2005.

2. How much more (in terms of percent) did a high-school graduate make than a non-high-school graduate in 1995?

3. How much more (in terms of percent) did a college graduate with a bachelor's degree make than a high-school graduate in 2005?

4. How much more (in terms of percent) did a person who had attended some college (without receiving a bachelor's degree) make than a high-school graduate in 2005?

(continued)

21. How Much Is School Worth?

5. Based on 2005 amounts, how much will each of the following make in 40 years in the workforce?

   a. a non-high-school graduate

   b. a high-school graduate

   c. a person with a bachelor's degree

6. Based on 2005 amounts, how much more can a person with a bachelor's degree expect to earn over 40 years than a high-school graduate? Express your answer as a dollar amount and a percent.

7. From 1985 to 2005, what was the percent increase in earnings for a person with a bachelor's degree? What annual increase does this represent over the 20-year period?

8. Assuming this kind of annual growth continues, predict the average salary for a person with a bachelor's degree the year you would graduate from college.

Percent of Increase/Decrease

teacher's page

22. Bigger, Stronger, and Faster

Context

sports

Topic

percent of increase/decrease

Overview

In this activity, students explore the concept of percent of increase and decrease by examining changes in Olympic records.

Objective

Students will be able to:

- find the percent of increase and decrease

Materials

- one copy of the Activity 22 handout for each student

Teaching Notes

- This activity works best for individual students.

- Some students may have difficulty working with the problems that involve distances. One strategy might be to change the distances to a common measurement unit such as feet or inches. Likewise, students may have trouble working with the times that involve both minutes and seconds. Again, have students work in minutes, or seconds if they are struggling.

Answers

1. 9.0%

2. 1900–1936, 6.4%

3. 7.35%

4. Since the trend shows both increases and decreases, it would be difficult to predict future distances.

5. 32.1% decrease

6. 2 minutes, 49.6 seconds

7. 6 feet, 5 inches, or almost 2.9% farther than her own throw

Extension Activity

Have students investigate the records in their favorite sports and determine percent of increase or decrease to make predictions about future records.

© 2007 Walch Publishing   Real-Life Math: Decimals and Percents

Name _____   Date _____

22. Bigger, Stronger, and Faster

If you follow the world of sports, you know that records are broken all the time. One way to observe these changes is to look at how distance records have increased and time records have decreased over time. The tables below list the winning times and distances for selected events and years in the Olympics. Use the information in each table to answer the questions that follow.

Men's 100-Meter Run

Year	Name	Country	Time (sec.)
1900	Francis W. Jarvis	United States	11.0
1936	Jesse Owens	United States	10.3
1964	Bob Hayes	United States	10.0
1988	Carl Lewis	United States	9.92
1996	Donovan Bailey	Canada	9.84
2004	Justin Gatlin	United States	9.85

Source: www.databaseolympics.com

1. By what percent did the winning times decrease from 1900 to 1964?

2. During which interval of time was the percent of decrease for the winning times the greatest? List the time interval and the percent.

Men's Long Jump

Year	Name	Country	Distance
1900	Alvin Kraenzlein	United States	23 ft., 6.75 in.
1936	Jesse Owens	United States	26 ft., 5.5 in.
1968	Bob Beamon	United States	29 ft., 2.5 in.
1988	Carl Lewis	United States	28 ft., 7.25 in.
1996	Carl Lewis	United States	27 ft., .75 in.
2004	Dwite Phillips	United States	28 ft., 2.2 in.

Source: www.databaseolympics.com

3. By what percent did the winning distance increase or decrease from 1968 to 1996?

(continued)

22. Bigger, Stronger, and Faster

4. Explain whether you would expect the winning distance in the 2012 Olympics to increase or decrease, and by what percent.

Women's 400-Meter Freestyle

Year	Name	Country	Time
1924	Martha Norelius	United States	6:02.2
1960	Susan "Chris" von Saltza	United States	4:50.6
1988	Janet Evans	United States	4:03.85
1996	Michelle Smith	Ireland	4:07.25
2000	Brooke Bennett	United States	4:05.8

Source: www.databaseolympics.com

5. From 1924 to 2000, by what percent did the winning time for the women's 400-meter freestyle decrease?

6. Starting in 2000, if the winning time decreased by the same percent over the next 76 years, what would the winning time be in the 2076 Olympics?

Women's Javelin Throw

Year	Name	Country	Distance
1932	"Babe" Didrikson	United States	143 ft., 4 in.
1996	Heli Rantanen	Finland	222 ft., 11 in.

Source: www.databaseolympics.com

7. How much farther would Heli Rantanen have to throw the javelin to increase her distance over Babe Didrikson's to 60%?

Percent of Increase/Decrease

teacher's page

23. The Ad Campaign

Context

travel/transportation

Topic

percent of increase/decrease

Overview

In this activity, students use percents to design an advertisement.

Objective

Students will be able to:

- find the percent of increase and decrease

Materials

- one copy of the Activity 23 handout for each student
- poster paper
- magazines with car ads (optional)

Teaching Notes

- Students can work individually, in pairs, or in small groups.
- As an introduction to the activity, have students look through magazines, paying attention to car advertisements and how numbers are used in them.
- Some students may need an explanation about the different categories found in the table.
- Emphasize with students the need to find the percent of increase or decrease prior to designing the advertisement.

Answers

Students' designs will vary. However, most advertisements should contain a significant number of categories.

1. The Ford costs 3.8% more.
2. The Chevy has a smaller (better) turning radius by 11.1%.
3. The Chevy gets better highway gas mileage by 9.4%.
4. The Chevy gets better city gas mileage by 4.5%.
5. The Ford has a bigger gas tank by 9.4%.
6. The Ford is 1% longer.
7. The Ford is 3.3% wider.
8. The Ford has a 1% longer wheelbase.
9. The Chevy has 3% more front headroom.
10. The Ford has 0.5% more rear headroom.
11. The Ford has 1% more front legroom.
12. The Chevy has 3.9% more rear legroom.
13. The Ford has 2.6% more maximum luggage capacity.

Extension Activity

Have students use desktop publishing software to create their advertisements.

23. The Ad Campaign

Imagine Chevrolet has just awarded your advertising agency a multimillion-dollar contract to design ads that will run in major magazines. As an up-and-coming ad designer, you have been assigned to work on the 2006 Chevy Malibu campaign. Your mission is to show that the Malibu is superior to the 2006 Ford Fusion by demonstrating in your ad that it has better specifications—it has more room, is less expensive, and so forth. In other words, your ad should point out the areas where the Malibu is better than the Fusion. Your supervisor wants you to identify differences using percents, since she feels that more people will understand the differences this way. Use the information below to design your ad.

For each category in the table, calculate the percent difference between the Chevy Malibu and the Ford Fusion.

Specifications for the 2006 Chevy Malibu and 2006 Ford Fusion

Category	Chevy Malibu	Ford Fusion	Percent difference
1. Price	$19,865	$20,625	
2. Turning Circle	36 feet	40 feet	
3. Highway miles per gallon	32 mpg	29 mpg	
4. City miles per gallon	22 mpg	21 mpg	
5. Fuel tank capacity	16 gallons	17.5 gallons	
6. Length	188.3 inches	190.2 inches	
7. Width	69.9 inches	72.2 inches	
8. Wheelbase	106.3 inches	107.4 inches	
9. Front headroom	39.9 inches	38.7 inches	
10. Rear headroom	37.6 inches	37.8 inches	
11. Front legroom	41.9 inches	42.3 inches	
12. Rear legroom	38.5 inches	37 inches	
13. Maximum luggage capacity	15.4 cubic feet	15.8 cubic feet	

Source: www.automotive.com and *Consumer Reports*

Percent of Increase/Decrease

teacher's page

24. CDs

Context

entertainment

Topic

percent of increase/decrease

Overview

In this activity, students make forecasts about manufacturing levels using percent of increase and decrease.

Objectives

Students will be able to:

- find the percent of increase and decrease
- use available information to make accurate predictions

Materials

- one copy of the Activity 24 handout for each student

Teaching Notes

- Students can work individually, in pairs, or in small groups.
- In the table, students can indicate which percent changes represent decreases by using a negative sign in front of the percent or brackets like these: < >.
- Some students may need help making the connection between knowing data trends and making predictions.

Answers

See the table below. Other answers will vary depending on students' predictions.

Extension Activity

Extend the production forecasts over the next several years.

CD	2002	2003	% change 2002–2003	2004	% change 2003–2004	2005	% change 2004–2005	2006	% change 2005–2006
Type 1	407.5	495.4	21.57	662.1	33.65	722.9	9.18	778.9	7.75
Type 2	366.4	339.5	–7.34	345.4	1.74	272.6	–21.08	225.3	–17.35
Type 3	2.3	1.2	–48.7	1.9	58.33	2.2	15.79	2.9	31.82

© 2007 Walch Publishing   *Real-Life Math: Decimals and Percents*

24. CDs

The company you work for manufactures three different types of recordable CDs. As a production supervisor, you have to forecast production levels for next year. To do that, you will examine the manufacturing trends over the previous years. The table below lists the production levels in millions for each of the three types of CD your company manufactured, and the amounts that were produced. Use that information to answer the questions that follow.

Units Manufactured, 2002–2006

CD	2002	2003	% change 2002–2003	2004	% change 2003–2004	2005	% change 2004–2005	200	% change 2005–2006
Type 1	407.5	495.4		662.1		722.9		778.9	
Type 2	366.4	339.5		345.4		272.6		225.3	
Type 3	2.3	1.2		1.9		2.2		2.9	

1. Fill in the table above with the percent changes from year to year.

2. How many units of each type of CD will you recommend be manufactured next year?

    type 1: _____

    type 2: _____

    type 3: _____

3. In the space below or on a separate sheet of paper, draft a memo to the production division manager that details your recommended production levels for the next year.

Problem Solving with Decimals and Percents

teacher's page

25. GPA

Context

sports

Topic

problem solving with decimals and percents

Overview

In this activity, students have to calculate what score they will need on their final exam in order to make the Conference Academic Team.

Objective

Students will be able to:

- solve problems involving decimals and percents

Materials

- one copy of the Activity 25 handout for each student

Teaching Notes

- This activity works best for individual students.

- As an introduction to the activity, ascertain from your students if they are aware of what grading method you use.

- Even if your grading policy is different from the two methods identified in the activity, most students will be familiar with the grading procedures mentioned in the activity.

Answers

Point Scale

1. homework: 82.5

   quizzes: 87

   chapter tests: 81

   current overall: 83.5

2. 189 points

Weighted Grades

1. homework: 12.94

   quizzes: 17.47

   chapter tests: 32.62

   total points: 63.03

2. 21.97

3. 88

Extension Activity

Allow students the opportunity to compute their current grade and set goals for future grades.

25. GPA

Imagine two of the goals you set for this year are to do well in sports and make the Conference Academic Team. In order to qualify for this recognition, you will have to earn an overall grade point average (GPA) of 85%. Let's say it is close to the end of the semester and the only class you're worried about is math. All the grades for the class are in, but you haven't taken the final exam yet. Use the information below to calculate what score you will need on the final exam to earn an 85 using two common grading methods, point scale and weighted grades.

Point Scale

Under this grading method, your math teacher assigns a certain number of points for each graded assignment or test. At the end of the semester, your grade is determined by finding the percent of points earned out of the total possible points. Listed in the table below are the points you have earned so far and the number of points possible for each category.

Grades to Date

Homework	Quizzes	Chapter Tests
99/120	174/200	405/500

1. Calculate your current grade in the class.

    homework grade: _____

    quiz grade: _____

    chapter test grade: _____

    current overall grade: _____

2. If the final exam is worth 200 points, what score will you need to get in order to have an overall average of 85% in the class? (*Hint:* To start the problem, add up the total number of points possible in the class and multiply by 85%.)

    score needed on final exam: _____

(continued)

Name _____ Date _____

25. GPA

Weighted Grades

Under this grading method, your math teacher uses the following weighted scale in determining grades. Homework is worth 15% of your grade, quizzes 20%, chapter tests 40%, and the final exam 25%. The table lists your grades to date.

Grades to Date

Homework	Quizzes	Chapter tests
86.25	87.34	81.56

1. First, you have to find out where you stand. Calculate your current grade using the information in the table.

   Example: If your homework average is 85, and homework is worth 15% of your grade, then change 15% to a decimal, 0.15, and multiply by your homework average, 85. Repeat this process for quizzes and chapter tests, and add the total points together.

   Homework = 0.15 × 85 = 12.75 points

   homework: _____

   quizzes: _____

   chapter tests: _____

   total points: _____

2. Since your goal in the course is to get an 85, find the difference between how many points you have and 85.

   points needed: _____

3. Next, figure the score you need to get on the final exam in order to earn an 85 in the course. Do this by dividing the number of points you need by the percent assigned to the final exam.

   score needed on final exam: _____

Problem Solving with Decimals and Percents

teacher's page

26. Decisions, Decisions

Context

money

Topic

problem solving with decimals and percents

Overview

In this activity, students are the owners of a small retail business and are in the process of deciding how they will pay their salespeople.

Objective

Students will be able to:

- solve problems involving decimals and percents

Materials

- one copy of the Activity 26 handout for each student

Teaching Notes

- Students can work individually or in pairs.
- Some students will need further explanation about how commissions work.

Answers

Either method is acceptable. Look for in-depth analysis and justification of why the students selected the pay method that they did.

Extension Activity

Have students contact local retailers and find out how they pay their employees.

Name _____   Date _____

26. Decisions, Decisions

Being the owner of a small retail business involves making a lot of decisions. One of the most important ones is figuring out how you are going to pay your salespeople. Since you want to attract and retain the best possible employees, you have to offer a compensation package that is both fair for you and rewards your top salespeople. After researching the different ways salespeople are paid, you have narrowed down your pay options to two.

Evaluate each of the pay methods below for a typical week. Once you have evaluated the different methods, select which method you want to use for your business. You expect your salespeople to generate about $500 a week in gross sales, and to work 15 hours.

> **Option 1: Graduated Commission**
>
> Using this method, a salesperson's total pay is based on the percentage of sales. As sales increase, so does the commission or percent the salesperson keeps. The scale you would use is listed below.
>
> - 18% on the first $300 of gross sales
> - 22% on the next $300 of gross sales
> - 30% on gross sales above $600
>
> **Option 2: Hourly Wage plus Commission**
>
> Under this method a salesperson would receive an hourly wage of $6.50 plus a commission of 6% of sales.

Decision Time

In the space below or on a separate sheet of paper, explain which pay method you will use and why. Remember, your goal is to attract and retain the highest-caliber salespeople, but you must also make money.

Problem Solving with Decimals and Percents

teacher's page

27. Owner Satisfaction

Context

entertainment

Topic

problem solving with decimals and percents

Overview

In this activity, students develop an owner satisfaction survey for video games.

Objective

Students will be able to:

- solve problems involving decimals and percents

Materials

- one copy of the Activity 27 handout for each student

Teaching Notes

- Students can work in small groups.
- Most students have had experience playing video games. Make sure the students who aren't familiar with video games are grouped with some "experts."
- Likewise, most students should be able to come up with five or six titles for their survey. If they can't, brainstorm with the whole class and make up a list of titles.
- Students should evaluate at least five criteria and five different games.
- Let students decide on the ranking system they will use. Some will ask for percent satisfied; some may use a zero-to-ten-point scale.

Answers

Answers will vary depending on video games and criteria selected.

Extension Activities

- Have students make graphs using spreadsheet software.
- Make the last question more formal by asking for a brief or a memo that provides plans for improving the owner satisfaction of video games.

Name _____   Date _____

27. Owner Satisfaction

Companies often use customer surveys to find out if customers are satisfied with the products they buy. Of course, if no one at the company was able to interpret the results of the survey, the information wouldn't do them much good. Luckily, this video game company is able to rely on your expertise in mathematics to help them out.

Design an owner satisfaction survey to evaluate video games. As part of the evaluation, you might take into consideration things such as level of difficulty, realism, fun, action, and so forth. You should also include an overall average for the game. The survey should include at least five different video games. A partial sample survey questionnaire is shown below. Use this as a model to design your own survey.

Sample Survey Form

Video game	Action	Level of difficulty	Fun	Realism		

1. Distribute and collect the surveys and calculate the overall average for each game.

2. Create a graph showing the overall average score for each video game.

3. Based on the information you collected in your survey, describe the strengths and weaknesses of the video games you examined.

Problem Solving with Decimals and Percents

teacher's page

28. On-Time Arrivals

Context

travel/transportation

Topic

problem solving with decimals and percents

Overview

In this activity, students lead a team that is responsible for improving the service of an airline.

Objective

Students will be able to:

- solve problems using decimals and percents

Materials

- one copy of the Activity 28 handout for each student

Teaching Notes

- Students can work individually, in pairs, or in small groups.

- Students with limited experience flying may need further explanation about what on-time arrivals and baggage handling claims are about.

- Prior to the activity, you may need to review the concepts of mean and median.

Answers

Student presentations will be different but should include mean and median for each topic.

On-time arrivals
mean: 80.42%
median: 80.55%
American ranked third.

Baggage handling errors
mean: 4.297
median: 3.875
American ranked tied for seventh.

Consumer complaints
mean: 0.76
median: 0.725
American ranked seventh.

Extension Activity

Have students research data for previous years and include that in their reports.

Additional Resource

www.transtats.bts.gov/

28. On-Time Arrivals

Imagine you have been assigned to lead American Airlines' newly formed Service Improvement Team (SIT). This indicates that top management recognizes you as a manager with tremendous skills. The goal of the SIT is to improve service in three areas: on-time arrivals, baggage handling, and consumer complaints. As team leader, your first step will be to analyze data in order to become familiar with American's track record in these three areas. Next, you will organize the data in such a way that you can present it at the first SIT meeting. As part of your presentation at the first meeting, you will show the airlines' mean and median for each of the three categories, and where American Airlines is ranked. Use the information in the following tables to create your presentation for the first SIT meeting.

On-Time Arrivals

Percentage of Domestic Flights Arriving Within 15 Minutes of Schedule					
Airline	JetBlue	American	Southwest	Continental	Alaska
Rate	75.1%	82.1%	80.8%	78.1%	80.5%
Airline	Northwest	US Airways	Delta	Amer. West	United
Rate	80.6%	79.8%	80.1%	84.1%	83.0%

Source: U.S. Depart of Transportation—Bureau of Transportation Statistics

Baggage Handling

Mishandled Baggage Reports per 1000 Passengers					
Airline	JetBlue	American	Southwest	Continental	Alaska
Rate	2.68	4.48	4.48	3.80	3.48
Airline	Northwest	US Airways	Delta	Amer. West	United
Rate	3.88	7.85	5.04	3.87	3.41

Source: U.S. Depart of Transportation

Consumer Complaints

Consumer Complaints per 100,000 Passengers					
Airline	JetBlue	American	Southwest	Continental	Alaska
Rate	0.29	0.79	0.24	1.03	0.38
Airline	Northwest	US Airways	Delta	Amer. West	United
Rate	0.73	1.71	0.99	0.72	0.72

Source: U.S. Depart of Transportation

(continued)

Name _____   Date _____

28. On-Time Arrivals

Complete the chart below to help organize your airline information.

Sorted Airline Information

rank	On-time airline	rate	Baggage airline	rate	Complaints airline	rate
1						
2						
3						
4						
5						
6						
7						
8						
9						
10						
mean						
median						

More Problem Solving with Decimals and Percents

teacher's page

29. The Big Sale

Context

money

Topic

problem solving with decimals and percents

Overview

In this activity, students assume the role of a promotions manager for a record store chain and must evaluate options regarding an upcoming sale.

Objective

Students will be able to:

- solve problems involving decimals and percents

Materials

- one copy of the Activity 29 handout for each student

Teaching Notes

- Students can work individually, in pairs, or in small groups.
- As an introduction to the activity, ask students about the last time they bought something on sale.
- For each of the sales options, revenue figures are listed for CDs. Students will need to determine which have been calculated correctly and which contain errors. For those that contain errors, students should make the corrections.

Answers

Option 1: correct

Option 2: CD revenue = $19.49

Option 3: CD revenue = $22.49

Option 4: correct

Option 5: correct

1–2. Answers will vary.

Extension Activity

Have students look through the newspaper or online and find advertisements for sales.

29. The Big Sale

Being in charge of sales and promotions for a major record store chain is an important position. It involves a thorough understanding of mathematics. Consider this situation: One of your local record store managers has submitted a proposal to you for an upcoming sale. The manager is considering structuring the sale in five different ways. However, you know that this particular store manager sometimes makes errors in his calculations.

You will have to review his proposal carefully and then recommend which sales strategy has the most potential. As a matter of company policy, sales are designed to maximize company profits. At the same time, they should sound attractive to the buyer. Carefully review each proposed sale option. Correct any calculation errors, and then choose the best sale strategy for the company.

Note: CDs are usually priced at $14.99 each.

> **Option 1:** Buy one CD at full price, get $5.00 off the second one.
> CD revenue = $24.98
>
> **Option 2:** Buy two CDs, receive 35% off each.
> CD revenue = $19.79
>
> **Option 3:** Buy one CD at full price, get the second for 50% off.
> CD revenue = $21.78
>
> **Option 4:** Buy two CDs and receive 40% off the total purchase.
> CD revenue = $17.99
>
> **Option 5:** Save 50% off your purchase of two CDs. Receive 25% off the first CD, and then save an additional 25% off that price on the second CD.
> CD revenue = $19.67

1. Choose the sale option that you will recommend. In the space below or on a separate sheet of paper, explain why you chose that option.

2. Make up another option for the sale. Include revenue figures.

More Problem Solving with Decimals and Percents

teacher's page

30. Will It Ever Fit?

Context

travel/transportation

Topic

problem solving with decimals and percents

Overview

In this activity, students assume the role of a logistics officer in the armed forces who is told to put together a plan for loading equipment and troops onto aircraft for an emergency deployment.

Objective

Students will be able to:

- solve problems involving decimals and percents

Materials

- one copy of the Activity 30 handout for each student
- manipulatives to represent the different types of equipment and vehicles: 25 objects total, split into four different color groups of 18, 4, 2, and 1; one set of 25 manipulatives for each group. (Algebra tiles are one possibility. You can also use manipulatives such as base 10 blocks to represent the troops.)

Teaching Notes

- Students can work in small groups.
- Explain to students that part of a logistics officer's job is to organize available resources to get people and equipment to the proper destination.
- It might help students solve this problem if they convert the payload capacities and weights from tons to pounds.
- Have students make a key for the manipulatives to help keep the information organized.
- Some groups will find it helpful to use four different sheets of paper to represent each aircraft, letting them experiment with different loading configurations more easily.
- The M1A2 tanks will only fit in the C-5 and C-17, not in a C-141.

Answers

Each loading plan will be different.

Extension Activity

Logisticians can be found working outside of the military at almost any major company. Have students contact a company and find out more about how logisticians use math in their jobs.

30. Will It Ever Fit?

Imagine this: You're the battalion logistics officer having your morning cup of coffee. Your commanding officer comes rushing into your office and tells you that you have until 1700 hours (5:00 P.M.) to put together a load-out plan for an emergency deployment. At first it sounds next to impossible. After thinking about it, you know that you can put the plan together by relying on your experience and vast knowledge of mathematics. Your first step is to find out how much equipment and how many troops you will have to find space for. Then you will need to find out what aircraft are available. You make a few phone calls and record the necessary information.

The following personnel and equipment are to be airlifted:

- Troops: 450 total (not including aircraft crew), 120 of whom must travel on the same plane

- Equipment: (2) M1A2 Abrams main battle tanks, (4) M3 Bradley fighting vehicles, (18) M998 cargo/troop carriers (Humvees), and miscellaneous support equipment

- Transport aircraft available: (1) C-5/B Galaxy, (2) C-17 Globemaster III, and (1) C-141B Starlifter

As you begin to draw up your loading plan, you realize that you need to consult your loading manuals to find out how much each aircraft can carry (payload) and how much each piece of equipment weighs.

Aircraft Payload Characteristics

	C-5	C-17	C-141
Maximum payload	102.452 tons	85.45 tons	44.85 tons
Maximum number of troops	73	102	200

Source: United States Air Force

Vehicle and Equipment Weights

M1A2 Abrams main battle tank	M3 Bradley fighting vehicle	Humvee	Miscellaneous equipment
69.54 tons	26.208 tons	3.1 tons	14.57 tons

Source: United States Army

(continued)

Name _____   Date _____

30. Will It Ever Fit?

Now that you have gathered the information you need, figure out how you are going to arrange the troops and equipment so they fit in the planes.

After you finalize your plan, complete the chart below to help organize the information.

Load-Out Plan

	C-5	C-17	C-17	C-141
Troops				
Equipment/Vehicles				

More Problem Solving with Decimals and Percents

teacher's page

31. How Many Calories Can I Eat?

Context

sports

Topic

problem solving with decimals and percents

Overview

In this activity, students calculate their daily calorie requirements and observe how those requirements change as they age.

Objective

Students will be able to:

- solve problems involving decimals and percents

Materials

- one copy of the Activity 31 handout for each student
- tape measures, one per group

Teaching Notes

- Students can work in pairs or small groups. If some students are sensitive about their weight, have all students work individually, or assign weights to students.

- Prior to the activity, ask students to find out how much they weigh, or borrow a scale from the gym and have students weigh themselves in class.

- You can avoid embarrassing students who are sensitive about their weight by not emphasizing losing weight during the activity.

- Have students measure their height in inches and convert to centimeters.

- You may need to help students decide on their level of activity. Question 6 allows students the opportunity to explore how different levels of activities affect calorie requirements.

Answers

Answers will vary depending on students' ages, weights, and heights.

Extension Activities

- Use a graphing calculator to model the equations.
- Graph the calorie requirements using a spreadsheet program.

Name _____   Date _____

31. How Many Calories Can I Eat?

For some of you, being a teenager means that you can eat and eat without worrying about gaining weight. As you get older, however, your body's metabolism slows down. This means your body requires fewer and fewer calories to maintain its present weight. In other words, to stay at the same weight, you will have to eat fewer calories. On the up side, you can figure out just how many calories you can eat as you age and still maintain your desired weight. Follow the steps below to figure it out.

1. Measure your height in centimeters. (1 inch = 2.5 centimeters)

    height (cm): _____

2. List your weight in kilograms. (2.2 pounds = 1 kilogram)

    weight (kg): _____

3. Use the guidelines below to determine your level of activity.

    - If you spend most of your time reading, sitting, eating, driving, and doing other sedentary activities, then use 1.1 for your activity level.

    - If you expend a moderate amount of energy doing things such as swimming, running, biking, or playing basketball, then use 1.5 for your activity level.

    - If you train competitively for triathlons or do activities such as boxing, long-distance running, and so forth, then use 2.0 for your activity level.

    - You can also select an appropriate amount in between the different levels.

    activity level: _____

4. Use the correct formula to determine your daily calorie requirements.

    males:

    [{66 + (weight in kg × 13.7) + (height in cm × 5) − (age × 6.8)} × activity level]

    females:

    [{655 + (weight in kg × 9.6) + (height in cm × 1.85) − (age × 4.7)} × activity level]

    daily calorie requirements: _____

5. As you get older, your metabolism slows down. This means your body needs fewer calories to maintain its current weight. Fill in the table on the next page with your calorie requirements for the given age. Assume that you will want to stay at whatever weight you are now. After you complete the table, use that information to create a graph that shows the relationship between age and calorie requirements.

(continued)

Name _____ Date _____

31. How Many Calories Can I Eat?

Daily Calorie Requirements

Age	15	20	25	30	35
Calories					
Age	40	45	50	55	60
Calories					

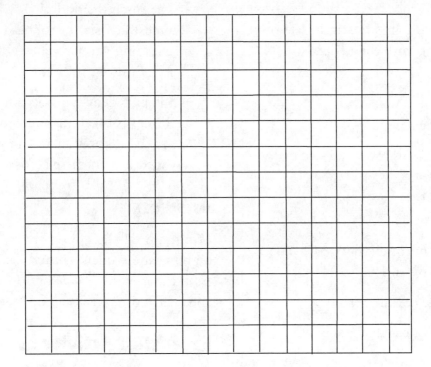

6. Find out how your activity level affects your calorie requirements by selecting a different level of activity and calculating calorie needs.

7. If there are 3500 calories in a pound, how might you use the information you found to gain or lose weight?

© 2007 Walch Publishing   Real-Life Math: Decimals and Percents

More Problem Solving with Decimals and Percents

teacher's page

32. Bungee!

Context
entertainment

Topic
problem solving with decimals and percents

Overview
In this activity, students use their math skills to make sure it is relatively safe to go bungee jumping with their friends.

Objective
Students will be able to:
- solve problems involving decimals and percents

Materials
- one copy of the Activity 32 handout for each student
- rubber bands, one per group
- paper clips, one per group
- weights, a variety of sizes; heavy metal washers or fishing weights work well

Teaching Notes
- Students can work in small groups.
- Introduce the activity by finding out if anyone in the class has ever gone bungee jumping. Elicit from those students how they think the bungee operators determined the safe height to place the cords.
- Bungee cords can be made in any size. The industry standard is 30 feet.
- The in-class simulation works with almost any size of rubber band, assuming you have a weight that will stretch it out fully.
- Some students will be concerned about the cords breaking because of weight. Let them know that weight is not a concern with these cords.

Answers
Minimum safe heights will vary depending on how far a student thinks is a safe distance from the ground: 250% of 30 feet is 75 feet, and 315% of 30 feet is 94.5 feet.

Extension Activity
Have students investigate the mathematics behind amusement park rides.

Name _____   Date _____

32. Bungee!

Imagine your friends have invited you to go bungee jumping with them this weekend, and for some reason, you have agreed. But you would like a few more details, especially about the bungee cord. Your friends tell you that it is a standard 30-foot cord with an elongation factor of 250% to 315%. Not that you don't trust your friends' mathematical ability, but you will feel a lot better if you calculate how high the bridge needs to be yourself. Before you go jumping, answer the questions that follow.

1. Using a 30-foot cord with the previously mentioned characteristics, what is the minimum height that you can safely jump from? Remember, you will jump headfirst and must take your own height into consideration.

    minimum safe height: _____

2. Suppose you have already chosen the ideal location for making a jump. Now all you have to do is figure out how long a bungee cord to order. Choose a suitable local spot, find out how high it is, and then figure out the longest bungee cord that you could use there.

    location and height: _____

    length of bungee cord: _____

Bungee Jump Simulation

Follow these steps to simulate a bungee jump in the classroom.

1. Cut a rubber band to produce a straight strip of elastic.

2. Lay the rubber band flat on a yardstick or meterstick and measure its length.

    length: _____

3. Stretch the rubber band to its full length and measure it.

    stretched length: _____

4. Calculate the elongation factor for your rubber band.

    elongation factor: _____

5. Tie a paper clip to one end of the rubber band and find a suitable spot to make a "jump."

6. Attach a weight to the paper clip and conduct a "jump."

© 2007 Walch Publishing   Real-Life Math: Decimals and Percents

Real-Life Math: Decimals and Percents

Share Your Bright Ideas

We want to hear from you!

Your name_____Date_____

School name_____

School address_____

City_____State_____Zip_____Phone number (_____)_____

Grade level(s) taught_____Subject area(s) taught_____

Where did you purchase this publication?_____

In what month do you purchase a majority of your supplements?_____

What moneys were used to purchase this product?

    ___School supplemental budget   ___Federal/state funding   ___Personal

Please "grade" this Walch publication in the following areas:

Quality of service you received when purchasing	A	B	C	D
Ease of use	A	B	C	D
Quality of content	A	B	C	D
Page layout	A	B	C	D
Organization of material	A	B	C	D
Suitability for grade level	A	B	C	D
Instructional value	A	B	C	D

COMMENTS:_____

_____

What specific supplemental materials would help you meet your current—or future—instructional needs?

_____

Have you used other Walch publications? If so, which ones?_____

May we use your comments in upcoming communications? ___Yes ___No

Please **FAX** this completed form to **888-991-5755**, or mail it to

   Customer Service, Walch Publishing, P. O. Box 658, Portland, ME 04104-0658

We will send you a **FREE GIFT** in appreciation of your feedback. **THANK YOU!**